Learn

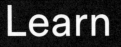

Eureka Math®
Grade 2
Module 8

Published by Great Minds®.

Copyright © 2018 Great Minds®.

Printed in the U.S.A.
This book may be purchased from the publisher at eureka-math.org.
BAB   10   9   8   7   6   5

ISBN 978-1-64054-058-3

G2-M8-L-05.2018

Learn ◆ Practice ◆ Succeed

Eureka Math® student materials for *A Story of Units*® (K–5) are available in the *Learn, Practice, Succeed* trio. This series supports differentiation and remediation while keeping student materials organized and accessible. Educators will find that the *Learn, Practice,* and *Succeed* series also offers coherent—and therefore, more effective—resources for Response to Intervention (RTI), extra practice, and summer learning.

Learn

Eureka Math Learn serves as a student's in-class companion where they show their thinking, share what they know, and watch their knowledge build every day. *Learn* assembles the daily classwork—Application Problems, Exit Tickets, Problem Sets, templates—in an easily stored and navigated volume.

Practice

Each *Eureka Math* lesson begins with a series of energetic, joyous fluency activities, including those found in *Eureka Math Practice*. Students who are fluent in their math facts can master more material more deeply. With *Practice,* students build competence in newly acquired skills and reinforce previous learning in preparation for the next lesson.

Together, *Learn* and *Practice* provide all the print materials students will use for their core math instruction.

Succeed

Eureka Math Succeed enables students to work individually toward mastery. These additional problem sets align lesson by lesson with classroom instruction, making them ideal for use as homework or extra practice. Each problem set is accompanied by a Homework Helper, a set of worked examples that illustrate how to solve similar problems.

Teachers and tutors can use *Succeed* books from prior grade levels as curriculum-consistent tools for filling gaps in foundational knowledge. Students will thrive and progress more quickly as familiar models facilitate connections to their current grade-level content.

Students, families, and educators:

Thank you for being part of the *Eureka Math®* community, where we celebrate the joy, wonder, and thrill of mathematics.

In the *Eureka Math* classroom, new learning is activated through rich experiences and dialogue. The *Learn* book puts in each student's hands the prompts and problem sequences they need to express and consolidate their learning in class.

What is in the Learn book?

Application Problems: Problem solving in a real-world context is a daily part of *Eureka Math*. Students build confidence and perseverance as they apply their knowledge in new and varied situations. The curriculum encourages students to use the RDW process—Read the problem, Draw to make sense of the problem, and Write an equation and a solution. Teachers facilitate as students share their work and explain their solution strategies to one another.

Problem Sets: A carefully sequenced Problem Set provides an in-class opportunity for independent work, with multiple entry points for differentiation. Teachers can use the Preparation and Customization process to select "Must Do" problems for each student. Some students will complete more problems than others; what is important is that all students have a 10-minute period to immediately exercise what they've learned, with light support from their teacher.

Students bring the Problem Set with them to the culminating point of each lesson: the Student Debrief. Here, students reflect with their peers and their teacher, articulating and consolidating what they wondered, noticed, and learned that day.

Exit Tickets: Students show their teacher what they know through their work on the daily Exit Ticket. This check for understanding provides the teacher with valuable real-time evidence of the efficacy of that day's instruction, giving critical insight into where to focus next.

Templates: From time to time, the Application Problem, Problem Set, or other classroom activity requires that students have their own copy of a picture, reusable model, or data set. Each of these templates is provided with the first lesson that requires it.

Where can I learn more about Eureka Math *resources?*

The Great Minds® team is committed to supporting students, families, and educators with an ever-growing library of resources, available at eureka-math.org. The website also offers inspiring stories of success in the *Eureka Math* community. Share your insights and accomplishments with fellow users by becoming a *Eureka Math* Champion.

Best wishes for a year filled with aha moments!

Jill Diniz

Jill Diniz
Director of Mathematics
Great Minds

The Read–Draw–Write Process

The *Eureka Math* curriculum supports students as they problem-solve by using a simple, repeatable process introduced by the teacher. The Read–Draw–Write (RDW) process calls for students to

1. Read the problem.

2. Draw and label.

3. Write an equation.

4. Write a word sentence (statement).

Educators are encouraged to scaffold the process by interjecting questions such as

- What do you see?

- Can you draw something?

- What conclusions can you make from your drawing?

The more students participate in reasoning through problems with this systematic, open approach, the more they internalize the thought process and apply it instinctively for years to come.

Contents

Module 8: Time, Shapes, and Fractions as Equal Parts of Shapes

R (Read the problem carefully.)

Terrence is making shapes with 12 toothpicks. Using all of the toothpicks, create 3 different shapes he could make. How many other combinations can you find?

D (Draw a picture.)

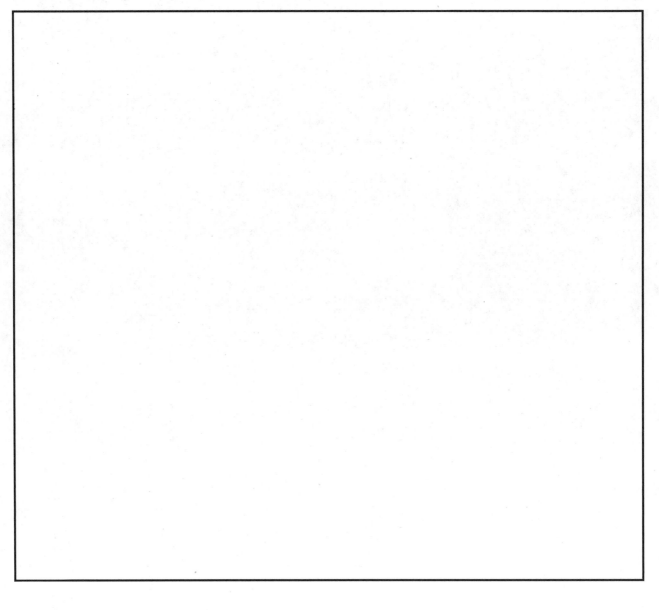

Name _____   Date _____

1. Identify the number of sides and angles for each shape. Circle each angle as you count, if needed. The first one has been done for you.

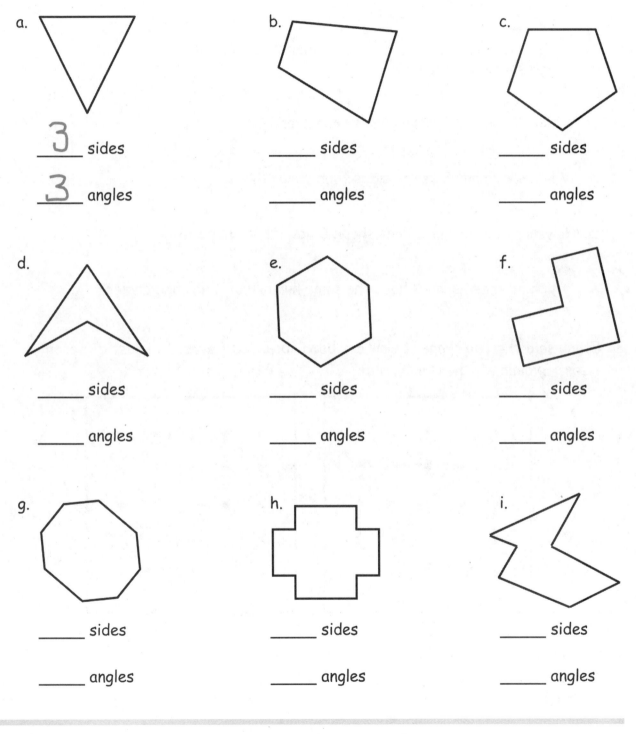

a.

__3__ sides

__3__ angles

b.

_____ sides

_____ angles

c.

_____ sides

_____ angles

d.

_____ sides

_____ angles

e.

_____ sides

_____ angles

f.

_____ sides

_____ angles

g.

_____ sides

_____ angles

h.

_____ sides

_____ angles

i.

_____ sides

_____ angles

2. Study the shapes below. Then, answer the questions.

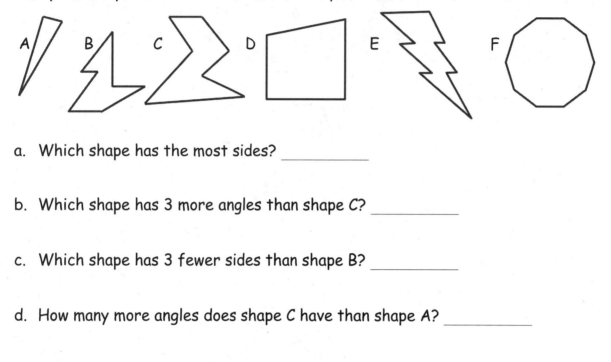

a. Which shape has the most sides? _____

b. Which shape has 3 more angles than shape C? _____

c. Which shape has 3 fewer sides than shape B? _____

d. How many more angles does shape C have than shape A? _____

e. Which of these shapes have the same number of sides and angles? _____

3. Ethan said the two shapes below are both six-sided figures but just different sizes. Explain why he is incorrect.

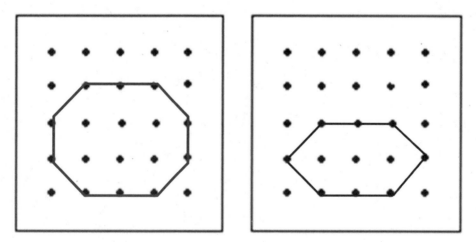

_____

_____

4   Lesson 1:   Describe two-dimensional shapes based on attributes.

© 2018 Great Minds®. eureka-math.org

EUREKA
MATH

Name _____   Date _____

Study the shapes below. Then, answer the questions.

A   B   C   D

1. Which shape has the most sides? _____

2. Which shape has 3 fewer angles than shape C? _____

3. Which shape has 3 more sides than shape B? _____

4. Which of these shapes have the same number of sides and angles? _____

R (Read the problem carefully.)

How many triangles can you find? (Hint: If you only found 10, keep looking!)

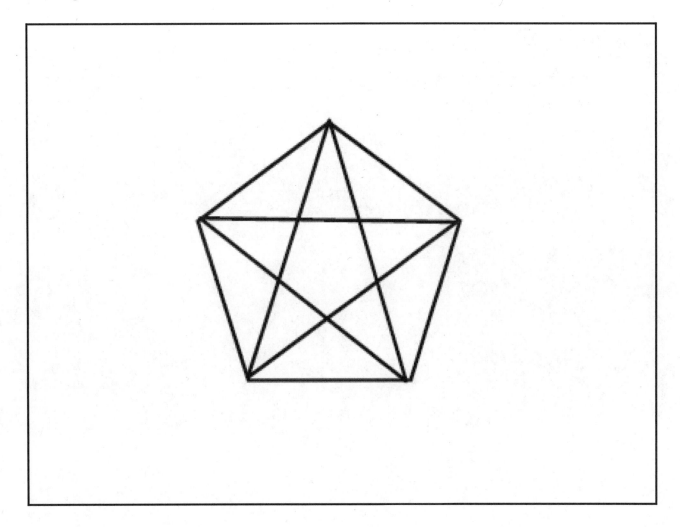

Lesson 2:   Build, identify, and analyze two-dimensional shapes with specified attributes.

© 2018 Great Minds®. eureka-math.org

7

W (Write a Statement that matches the story.)

_____

_____

_____

Lesson 2:   Build, identify, and analyze two-dimensional shapes with specified attributes.

© 2018 Great Minds®. eureka-math.org

EUREKA
MATH

Name _____   Date _____

1. Count the number of sides and angles for each shape to identify each polygon. The polygon names in the word bank may be used more than once.

| Hexagon | Quadrilateral | Triangle | Pentagon |

a.   b.   c.

_____   _____   _____

d.   e.   f.

_____   _____   _____

g.   h.   i.

_____   _____   _____

j.   k.   l.

_____   _____   _____

EUREKA MATH

Lesson 2:   Build, identify, and analyze two-dimensional shapes with specified attributes.

© 2018 Great Minds®. eureka-math.org

9

2. Draw more sides to complete 2 examples of each polygon.

	Example 1	Example 2
a. **Triangle** For each example, _____ line was added. A triangle has _____ total sides.		
b. **Hexagon** For each example, _____ lines were added. A hexagon has _____ total sides.		
c. **Quadrilateral** For each example, _____ lines were added. A quadrilateral has _____ total sides.		
d. **Pentagon** For each example, _____ lines were added. A pentagon has _____ total sides.		

3.

a. Explain why both polygons A and B are hexagons.

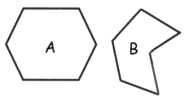

_____

_____

b. Draw a different hexagon than the two that are shown.

4. Explain why both polygons C and D are quadrilaterals.

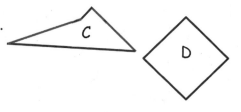

_____

_____

Lesson 2:   Build, identify, and analyze two-dimensional shapes with specified attributes.

© 2018 Great Minds®. eureka-math.org

Name _____   Date _____

Count the number of sides and angles for each shape to identify each polygon.
The polygon names in the word bank may be used more than once.

| Hexagon   Quadrilateral   Triangle   Pentagon |

1.

_____

2.

_____

3.

_____

4.

_____

5.

_____

6.

_____

Lesson 2:   Build, identify, and analyze two-dimensional shapes with specified
attributes.

© 2018 Great Minds®. eureka-math.org

11

R (Read the problem carefully.)

Three sides of a quadrilateral have the following lengths: 19 cm, 23 cm, and 26 cm. If the total distance around the shape is 86 cm, what is the length of the fourth side?

D (Draw a picture.)

W (Write and solve an equation.)

Lesson 3:   Use attributes to draw different polygons including triangles, quadrilaterals, pentagons, and hexagons.

© 2018 Great Minds®. eureka-math.org

13

W (Write a Statement that matches the story.)

_____

_____

_____

Lesson 3: Use attributes to draw different polygons including triangles, quadrilaterals, pentagons, and hexagons.

Name _____   Date _____

1. Use a straightedge to draw the polygon with the given attributes in the space to the right.

    a. Draw a polygon with 3 angles.

        Number of sides: _____

        Name of polygon: _____

    b. Draw a five-sided polygon.

        Number of angles: _____

        Name of polygon: _____

    c. Draw a polygon with 4 angles.

        Number of sides: _____

        Name of polygon: _____

    d. Draw a six-sided polygon.

        Number of angles: _____

        Name of polygon: _____

    e. Compare your polygons to those of your partner.

        Copy one example that is very different from your own in the space to the right.

Lesson 3:   Use attributes to draw different polygons including triangles,
               quadrilaterals, pentagons, and hexagons.

© 2018 Great Minds®. eureka-math.org

15

2. Use your straightedge to draw 2 new examples of each polygon that are different from those you drew on the first page.

a. Triangle

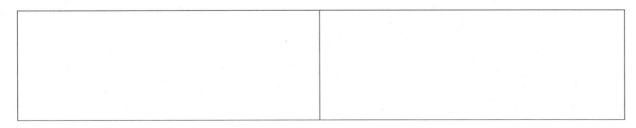

b. Pentagon

c. Quadrilateral

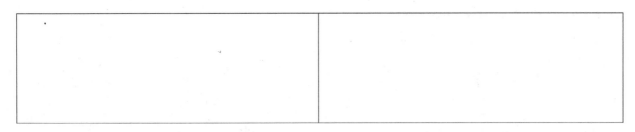

d. Hexagon

Lesson 3:   Use attributes to draw different polygons including triangles, quadrilaterals, pentagons, and hexagons.

© 2018 Great Minds®. eureka-math.org

Name _____   Date _____

Use a straightedge to draw the polygon with the given attributes in the space to the right.

Draw a five-sided polygon.

Number of angles: _____

Name of polygon: _____

Lesson 3:   Use attributes to draw different polygons including triangles,
quadrilaterals, pentagons, and hexagons.

© 2018 Great Minds®. eureka-math.org

17

Name _____   Date _____

1. Use your ruler to draw 2 parallel lines that are not the same length.

2. Use your ruler to draw 2 parallel lines that are the same length.

3. Trace the parallel lines on each quadrilateral using a crayon. For each shape with two sets of parallel lines, use two different colors. Use your index card to find each square corner, and box it.

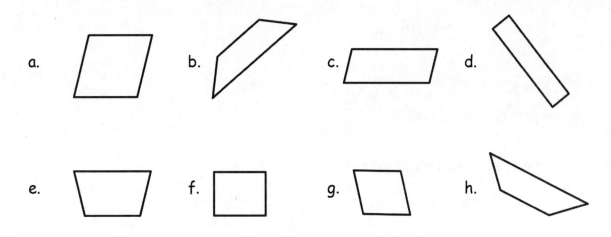

a.   b.   c.   d.

e.   f.   g.   h.

4. Draw a parallelogram with no square corners.

Lesson 4:   Use attributes to identify and draw different quadrialaterals including rectangles, rhombuses, parallelograms, and trapezoids.

© 2018 Great Minds®. eureka-math.org

19

5. Draw a quadrilateral with 4 square corners.

6. Measure and label the sides of the figure to the right with
    your centimeter ruler. What do you notice? Be ready to talk
    about the attributes of this quadrilateral. Can you
    remember what this polygon is called?

7. A square is a special rectangle. What makes it special?

    _____

    _____

Lesson 4:   Use attributes to identify and draw different quadrialaterals including
                      rectangles, rhombuses, parallelograms, and trapezoids.

© 2018 Great Minds®. eureka-math.org

EUREKA
MATH®

Name _____   Date _____

Use crayons to trace the parallel sides on each quadrilateral. Use your index card to find each square corner, and box it.

1.   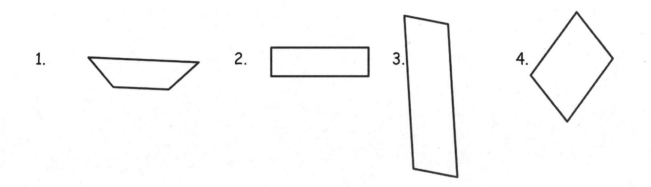   2.   3.   4.

Lesson 4: Use attributes to identify and draw different quadrilaterals including rectangles, rhombuses, parallelograms, and trapezoids.

© 2018 Great Minds®. eureka-math.org

21

R (Read the problem carefully.)

Owen had 90 straws to create pentagons. He created a set of 5 pentagons when he noticed a number pattern. How many more shapes can he add to the pattern?

D (Draw a picture.)

W (Write and solve an equation.)

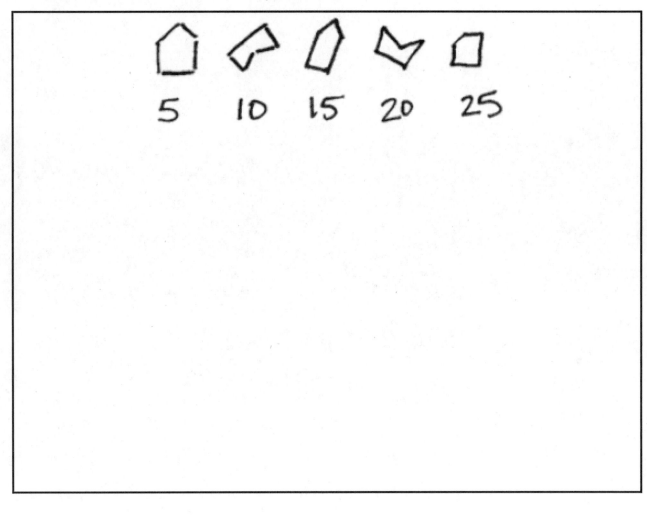

Lesson 5: Relate the square to the cube, and describe the cube based on attributes.

© 2018 Great Minds®. eureka-math.org

23

W (Write a Statement that matches the story.)

_____

_____

_____

Lesson 5:   Relate the square to the cube, and describe the cube based on attributes.

© 2018 Great Minds®. eureka-math.org

EUREKA MATH

Name _____   Date _____

1. Circle the shape that could be the face of a cube.

2. What is the most precise name of the shape you circled? _____

3. How many faces does a cube have? _____

4. How many edges does a cube have? _____

5. How many corners does a cube have? _____

6. Draw 6 cubes, and put a star next to your best one.

First cube	Second cube
Third cube	Fourth cube
Fifth cube	Sixth cube

Lesson 5:   Relate the square to the cube, and describe the cube based on attributes.

25

7. Connect the corners of the squares to make a different kind of drawing of a cube. The first one is done for you.

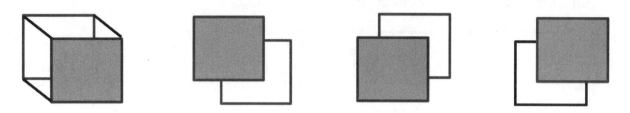

8. Derrick looked at the cube below. He said that a cube only has 3 faces. Explain why Derrick is incorrect.

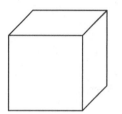

_____

_____

_____

Lesson 5:   Relate the square to the cube, and describe the cube based on attributes.

Name _____   Date _____

Draw 3 cubes. Put a star next to your best one.

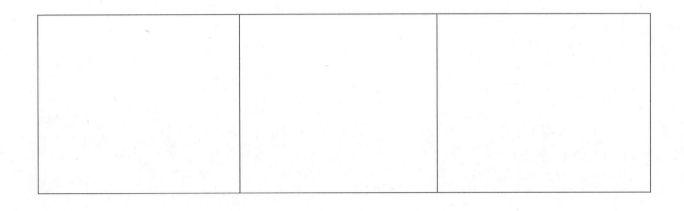

Lesson 5:   Relate the square to the cube, and describe the cube based on attributes.

© 2018 Great Minds®. eureka-math.org

27

R (Read the problem carefully.)

Frank has 19 fewer cubes than Josie. Frank has 56 cubes. They want to use all of their cubes to build a tower. How many cubes will they use?

D (Draw a picture.)

W (Write and solve an equation.)

Lesson 6:   Combine shapes to create a composite shape; create a new shape from composite shapes.

© 2018 Great Minds®. eureka-math.org

29

W (Write a statement that matches the story.)

_____

_____

_____

Lesson 6: Combine shapes to create a composite shape; create a new shape from composite shapes.

Name _____   Date _____

1. Identify each polygon labeled in the tangram as precisely as possible in the space below.

    a. _____

    b. _____

    c. _____

2. Use the square and the two smallest triangles of your tangram pieces to make the following polygons. Draw them in the space provided.

a. A quadrilateral with 1 pair of parallel sides.	b. A quadrilateral with no square corners.
c. A quadrilateral with 4 square corners.	d. A triangle with 1 square corner.

Lesson 6:   Combine shapes to create a composite shape; create a new shape from composite shapes.

31

© 2018 Great Minds®. eureka-math.org

3. Use the parallelogram and the two smallest triangles of your tangram pieces to make the following polygons. Draw them in the space provided.

a. A quadrilateral with 1 pair of parallel sides.	b. A quadrilateral with no square corners.
c. A quadrilateral with 4 square corners.	d. A triangle with 1 square corner.

4. Rearrange the parallelogram and the two smallest triangles to make a hexagon. Draw the new shape below.

5. Rearrange your tangram pieces to make other polygons! Identify them as you work.

Lesson 6:   Combine shapes to create a composite shape; create a new shape from composite shapes.

Name _____   Date _____

Use your tangram pieces to make two new polygons. Draw a picture of each new polygon, and name them.

Lesson 6:   Combine shapes to create a composite shape; create a new shape from composite shapes.

© 2018 Great Minds®. eureka-math.org

Cut the tangram into 7 puzzle pieces.

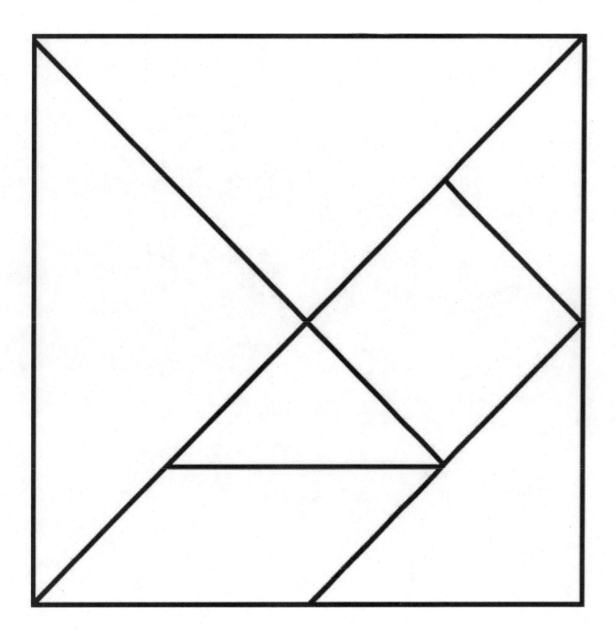

tangram

Lesson 6:   Combine shapes to create a composite shape; create a new shape
from composite shapes.

35

R (Read the problem carefully.)

Mrs. Libarian's students are picking up tangram pieces. The collect
13 parallelograms, 24 large triangles, 24 small triangles, and 13 medium
triangles. The rest are squares. If they collect 97 pieces in all, how many
squares are there?

D (Draw a picture.)

W (Write and solve an equation.)

Lesson 7:   Interpret equal shares in composite shapes as halves, thirds, and
             fourths.

© 2018 Great Minds®. eureka-math.org

37

W (Write a statement that matches the story.)

_____

_____

_____

Lesson 7:   Interpret equal shares in composite shapes as halves, thirds, and
             fourths.

Name _____ Date _____

1. Solve the following puzzles using your tangram pieces. Draw your solutions in the space below.

a. Use the two smallest triangles to make one larger triangle.	b. Use the two smallest triangles to make a parallelogram with no square corners.
c. Use the two smallest triangles to make a square.	d. Use the two largest triangles to make a square.
e. How many equal shares do the larger shapes in Parts (a–d) have?	f. How many halves make up the larger shapes in Parts (a–d)?

2. Circle the shapes that show halves.

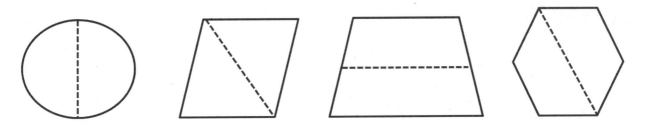

Lesson 7:   Interpret equal shares in composite shapes as halves, thirds, and fourths.

© 2018 Great Minds®. eureka-math.org

39

3. Show how 3 triangle pattern blocks form a trapezoid with one pair of parallel lines. Draw the shape below.

a. How many equal shares does the trapezoid have? _____

b. How many thirds are in the trapezoid? _____

4. Circle the shapes that show thirds.

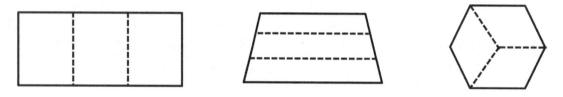

5. Add another triangle to the trapezoid you made in Problem 3 to make a parallelogram. Draw the new shape below.

a. How many equal shares does the shape have now? _____

b. How many fourths are in the shape? _____

6. Circle the shapes that show fourths.

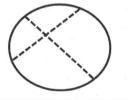 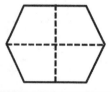 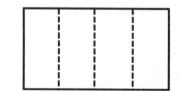

Lesson 7:   Interpret equal shares in composite shapes as halves, thirds, and fourths.

Name _____   Date _____

1. Circle the shapes that show thirds.

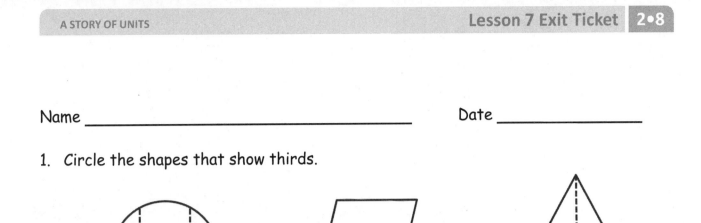

2. Circle the shapes that show fourths.

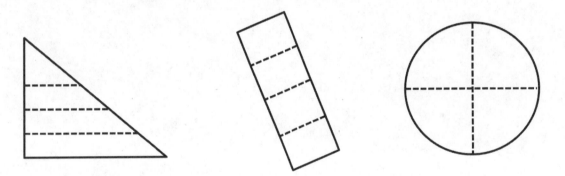

Lesson 7:   Interpret equal shares in composite shapes as halves, thirds, and fourths.

© 2018 Great Minds®. eureka-math.org

41

R (Read the problem carefully.)

Students were making larger shapes out of triangles and squares.

They put away all 72 triangles. There were still 48 squares on the carpet.

How many triangles and squares were on the carpet when they started?

D (Draw a picture.)

W (Write and solve an equation.)

EUREKA
MATH

Lesson 8: Interpret equal shares in composite shapes as halves, thirds, and fourths.

© 2018 Great Minds®. eureka-math.org

43

W (Write a statement that matches the story.)

_____

_____

_____

Lesson 8:   Interpret equal shares in composite shapes as halves, thirds, and
fourths.

© 2018 Great Minds®. eureka-math.org

Name _____   Date _____

1. Use one pattern block to cover half the rhombus.

    a. Identify the pattern block used to cover half of the rhombus. _____

    b. Draw a picture of the rhombus formed by the 2 halves.

2. Use one pattern block to cover half the hexagon.

    a. Identify the pattern block used to cover half of a hexagon. _____

    b. Draw a picture of the hexagon formed by the 2 halves.

3. Use one pattern block to cover 1 third of the hexagon.

    a. Identify the pattern block used to cover 1 third of a hexagon. _____

    b. Draw a picture of the hexagon formed by the 3 thirds.

4. Use one pattern block to cover 1 third of the trapezoid.

    a. Identify the pattern block used to cover 1 third of a trapezoid. _____

    b. Draw a picture of the trapezoid formed by the 3 thirds.

Lesson 8:   Interpret equal shares in composite shapes as halves, thirds, and
             fourths.

© 2018 Great Minds®. eureka-math.org

45

5. Use 4 pattern block squares to make one larger square.

    a. Draw a picture of the square formed in the space below.

    b. Shade 1 small square. Each small square is 1 _____ (half / third / fourth) of the whole square.

    c. Shade 1 more small square. Now, 2 _____ (halves / thirds / fourths) of the whole square is shaded.

    d. And 2 fourths of the square is the same as 1 _____ (half / third / fourth) of the whole square.

    e. Shade 2 more small squares. _____ fourths is equal to 1 whole.

6. Use one pattern block to cover 1 sixth of the hexagon.

    a. Identify the pattern block used to cover 1 sixth of a hexagon. _____

    b. Draw a picture of the hexagon formed by the 6 sixths.

Lesson 8:   Interpret equal shares in composite shapes as halves, thirds, and fourths.

© 2018 Great Minds®. eureka-math.org

Name _____   Date _____

Name the pattern block used to cover half the rectangle. _____

Use the shape below to draw the pattern blocks used to cover 2 halves.

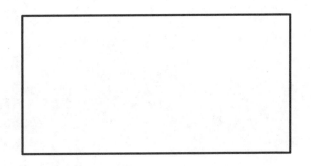

Lesson 8:   Interpret equal shares in composite shapes as halves, thirds, and fourths.

© 2018 Great Minds®. eureka-math.org

47

R (Read the problem carefully.)

Mr. Thompson's class raised 96 dollars for a field trip. They need to raise a total of 120 dollars.

    a. How much more money do they need to raise in order to reach their goal?

    b. If they raise 86 more dollars, how much extra money will they have?

D (Draw a picture.)

W (Write and solve an equation.)

Lesson 9:   Partition circles and rectangles into equal parts, and describe those parts as halves, thirds, or fourths.

© 2018 Great Minds®. eureka-math.org

49

W (Write a statement that matches the story.)

a. _____

_____

b. _____

_____

Lesson 9: Partition circles and rectangles into equal parts, and describe those parts as halves, thirds, or fourths.

EUREKA MATH

Name _____   Date _____

1. Circle the shapes that have 2 equal shares with 1 share shaded.

2. Shade 1 half of the shapes that are split into 2 equal shares. One has been done for you.

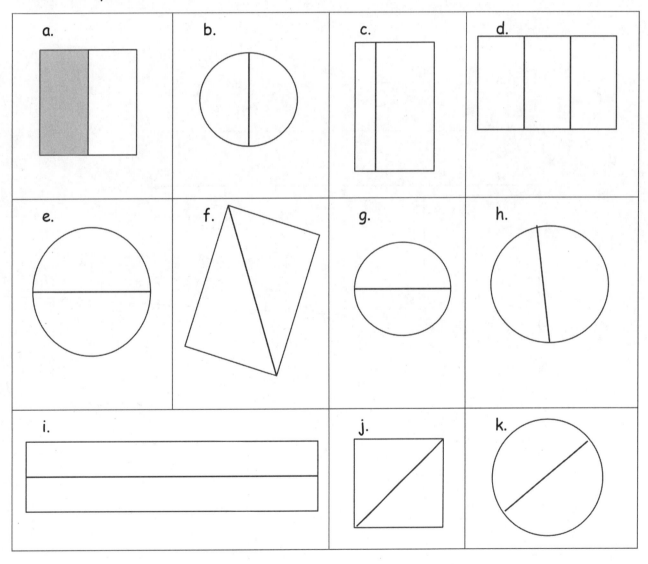

   Lesson 9:   Partition circles and rectangles into equal parts, and describe those
parts as halves, thirds, or fourths.

© 2018 Great Minds®. eureka-math.org

3. Partition the shapes to show halves. Shade 1 half of each. Compare your halves to your partner's.

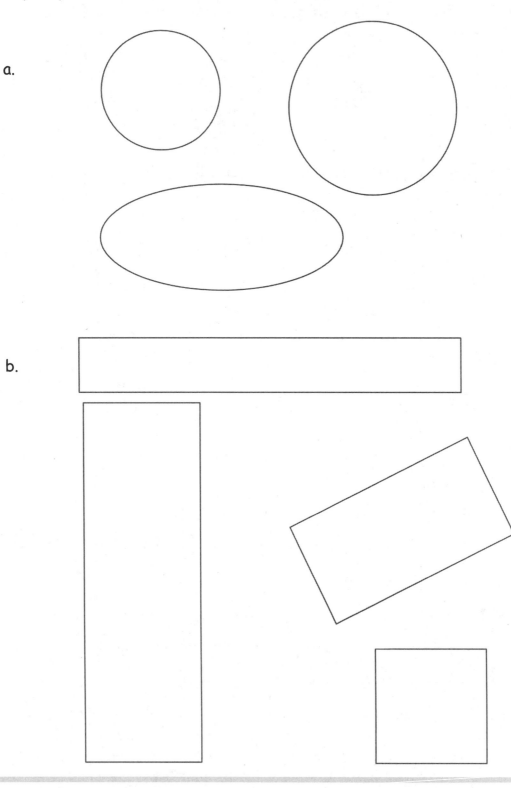

a.

b.

Lesson 9:   Partition circles and rectangles into equal parts, and describe those parts as halves, thirds, or fourths.

© 2018 Great Minds®. eureka-math.org

Name _____   Date _____

Shade 1 half of the shapes that are split into 2 equal shares.

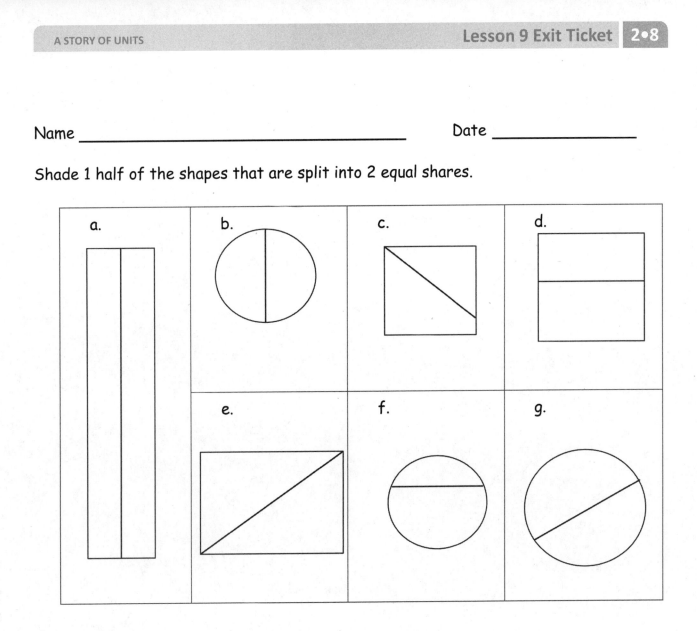

Lesson 9:   Partition circles and rectangles into equal parts, and describe those parts as halves, thirds, or fourths.

© 2018 Great Minds®. eureka-math.org

53

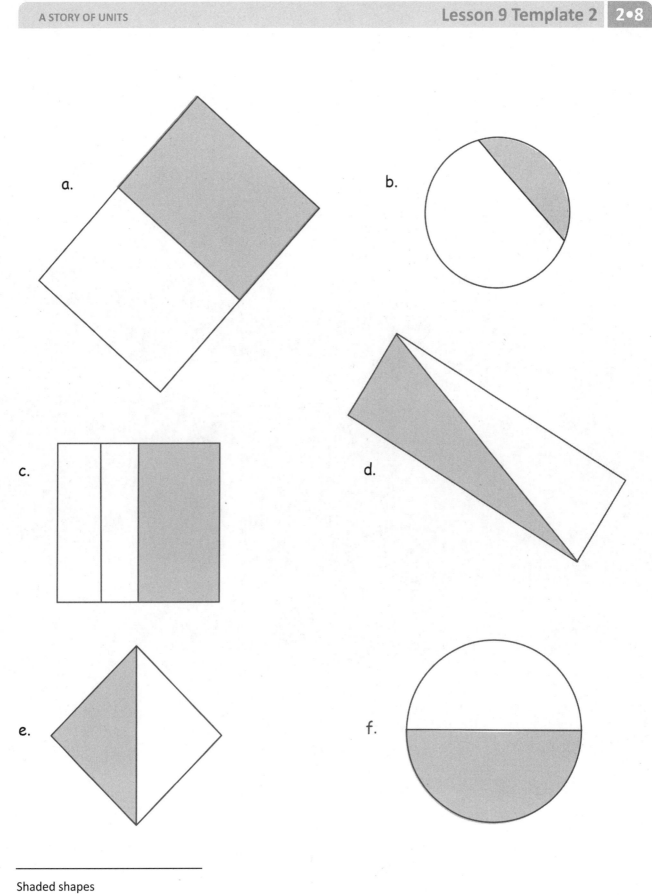

a.

b.

c.

d.

e.

f.

Shaded shapes

EUREKA MATH

Lesson 9: Partition circles and rectangles into equal parts, and describe those parts as halves, thirds, or fourths.

55

© 2018 Great Minds®. eureka-math.org

R (Read the problem carefully.)

Felix is passing out raffle tickets. He passes out 98 tickets and has 57 left. How many raffle tickets did he have to start?

D (Draw a picture.)

W (Write and solve an equation.)

EUREKA
MATH®

Lesson 10: Partition circles and rectangles into equal parts, and describe those parts as halves, thirds, or fourths.

© 2018 Great Minds®. eureka-math.org

57

W (Write a statement that matches the story.)

_____

_____

_____

Lesson 10:   Partition circles and rectangles into equal parts, and describe those
                            parts as halves, thirds, or fourths.

Name _____   Date _____

1. a. Do the shapes in Problem 1(a) show halves or thirds? _____

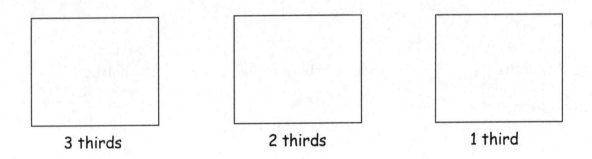

   b. Draw 1 more line to partition each shape above into fourths.

2. Partition each rectangle into thirds. Then, shade the shapes as indicated.

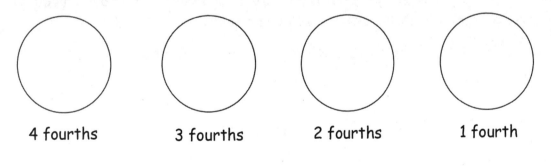

3 thirds   2 thirds   1 third

3. Partition each circle into fourths. Then, shade the shapes as indicated.

4 fourths   3 fourths   2 fourths   1 fourth

EUREKA MATH®

Lesson 10: Partition circles and rectangles into equal parts, and describe those parts as halves, thirds, or fourths.

© 2018 Great Minds®. eureka-math.org

59

4. Partition and shade the following shapes as indicated. Each rectangle or circle is one whole.

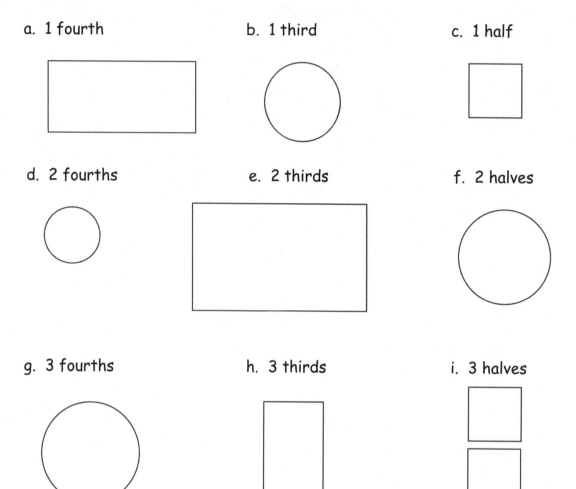

a. 1 fourth

b. 1 third

c. 1 half

d. 2 fourths

e. 2 thirds

f. 2 halves

g. 3 fourths

h. 3 thirds

i. 3 halves

5. Split the pizza below so that Maria, Paul, Jose, and Mark each have an equal share. Label each student's share with his or her name.

a. What fraction of the pizza was eaten by each of the boys?

b. What fraction of the pizza did the boys eat altogether?

Lesson 10: Partition circles and rectangles into equal parts, and describe those parts as halves, thirds, or fourths.

Name _____   Date _____

Partition and shade the following shapes as indicated. Each rectangle or circle is one whole.

1. 2 halves

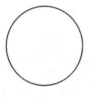

2. 2 thirds

3. 1 third

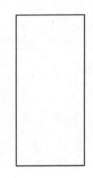

4. 1 half

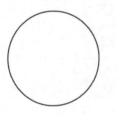

5. 2 fourths

6. 1 fourth

EUREKA MATH

Lesson 10:   Partition circles and rectangles into equal parts, and describe those parts as halves, thirds, or fourths.

© 2018 Great Minds®. eureka-math.org

61

rectangles and circles

EUREKA MATH®

Lesson 10:   Partition circles and rectangles into equal parts, and describe those parts as halves, thirds, or fourths.

63

© 2018 Great Minds®. eureka-math.org

R (Read the problem carefully.)

Jacob collected 70 baseball cards. He gave half of them to his brother, Sammy. How many baseball cards does Jacob have left?

D (Draw a picture.)

W (Write and solve an equation.)

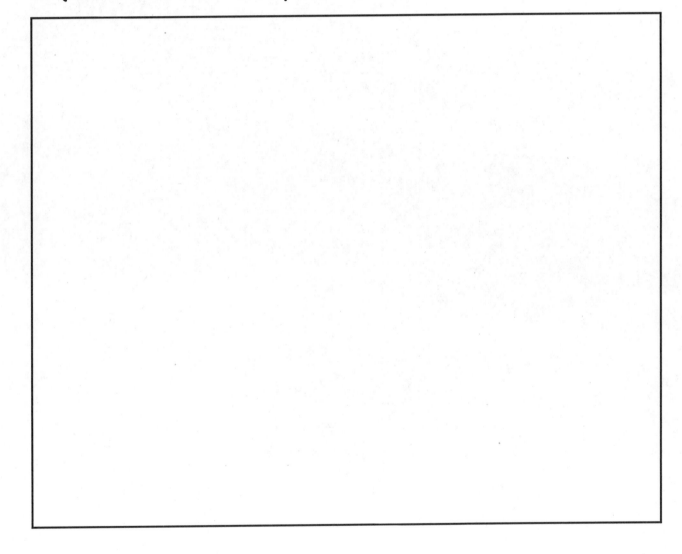

EUREKA MATH

Lesson 11:   Describe a whole by the number of equal parts including 2 halves, 3 thirds, and 4 fourths.

© 2018 Great Minds®. eureka-math.org

65

W (Write a statement that matches the story.)

_____

_____

_____

Lesson 11:   Describe a whole by the number of equal parts including 2 halves, 3 thirds, and 4 fourths.

© 2018 Great Minds®. eureka-math.org

Name _____   Date _____

1. For Parts (a), (c), and (e), identify the shaded area.

a.

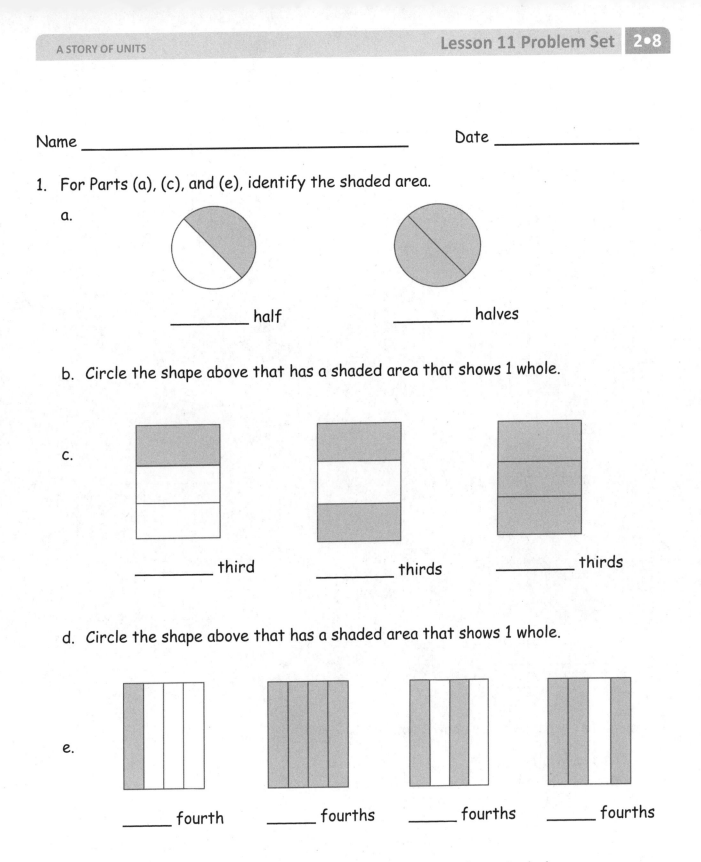

_____ half   _____ halves

b. Circle the shape above that has a shaded area that shows 1 whole.

c.

_____ third   _____ thirds   _____ thirds

d. Circle the shape above that has a shaded area that shows 1 whole.

e.

_____ fourth   _____ fourths   _____ fourths   _____ fourths

f. Circle the shape above that has a shaded area that shows 1 whole.

Lesson 11: Describe a whole by the number of equal parts including 2 halves,
3 thirds, and 4 fourths.

67

© 2018 Great Minds®. eureka-math.org

2. What fraction do you need to color so that 1 whole is shaded?

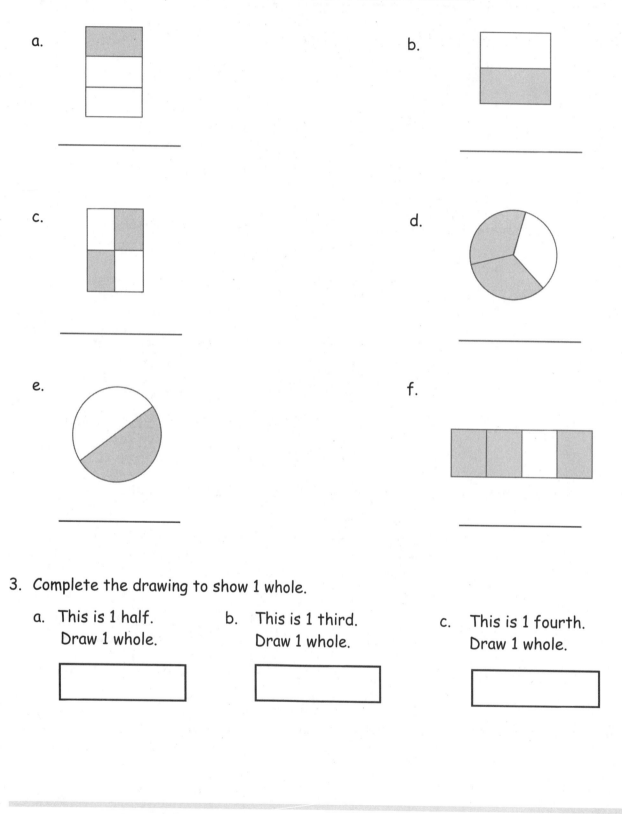

a. _____

b. _____

c. _____

d. _____

e. _____

f. _____

3. Complete the drawing to show 1 whole.

   a. This is 1 half.
       Draw 1 whole.

   b. This is 1 third.
       Draw 1 whole.

   c. This is 1 fourth.
       Draw 1 whole.

Lesson 11:   Describe a whole by the number of equal parts including 2 halves,
              3 thirds, and 4 fourths.

© 2018 Great Minds®. eureka-math.org

EUREKA
MATH®

Name _____   Date _____

What fraction do you need to color so that 1 whole is shaded?

1.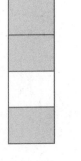

   _____

2.

   _____

3.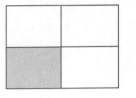

   _____

4.

   _____

Lesson 11:   Describe a whole by the number of equal parts including 2 halves,
3 thirds, and 4 fourths.

© 2018 Great Minds®. eureka-math.org

69

R (Read the problem carefully.)

Tugu made two pizzas for himself and his 5 friends to share. He wants everyone to have an equal share of the pizza. Should he cut the pizzas into halves, thirds, or fourths?

D (Draw a picture.)

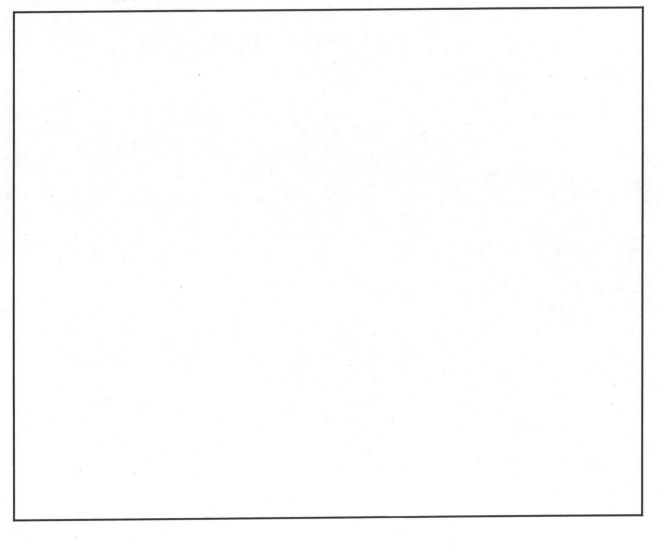

Lesson 12:   Recognize that equal parts of an identical rectangle can have different shapes.

© 2018 Great Minds®. eureka-math.org

71

W (Write a statement that matches the story.)

_____

_____

_____

Lesson 12: Recognize that equal parts of an identical rectangle can have different shapes.

© 2018 Great Minds®. eureka-math.org

Name _____   Date _____

1. Partition the rectangles in 2 different ways to show equal shares.

   a. 2 halves

   b. 3 thirds

   c. 4 fourths

2. Build the original whole square using the rectangle half and the half represented by your 4 small triangles. Draw it in the space below.

Lesson 12:   Recognize that equal parts of an identical rectangle can have different shapes.

© 2018 Great Minds®. eureka-math.org

73

3. Use different-colored halves of a whole square.

   a. Cut the square in half to make 2 equal-size rectangles.

   b. Rearrange the halves to create a new rectangle with no gaps or overlaps.

   c. Cut each equal part in half to make 4 equal-size squares.

   d. Rearrange the new equal shares to create different polygons.

   e. Draw one of your new polygons from Part (d) below.

Extension

4. Cut out the circle.

   a. Cut the circle in half.

   b. Rearrange the halves to create a new shape with no gaps or overlaps.

   c. Cut each equal share in half.

   d. Rearrange the equal shares to create a new shape with no gaps or overlaps.

   e. Draw your new shape from Part (d) below.

Lesson 12:   Recognize that equal parts of an identical rectangle can have different
                                shapes.

Name _____   Date _____

Partition the rectangles in 2 different ways to show equal shares.

1. 2 halves

2. 3 thirds

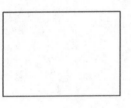 

3. 4 fourths

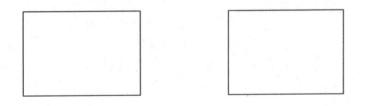

Lesson 12:   Recognize that equal parts of an identical rectangle can have different shapes.

© 2018 Great Minds®. eureka-math.org

75

Name _____   Date _____

1. Tell what fraction of each clock is shaded in the space below using the words
    quarter, quarters, half, or *halves.*

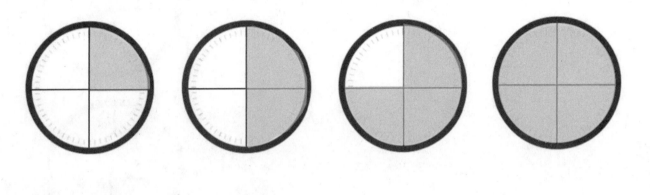

_____   _____   _____   _____

2. Write the time shown on each clock.

a.

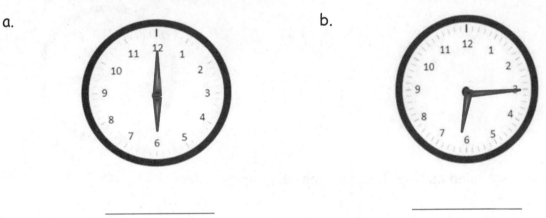

_____

b.

_____

c.

_____

d.

_____

Lesson 13:   Construct a paper clock by partitioning a circle into halves and
              quarters, and tell time to the half hour or quarter hour.

© 2018 Great Minds®. eureka-math.org

77

3. Match each time to the correct clock by drawing a line.

- Quarter to 4

- Half past 8

- 8:30

- 3:45

- 1:15

3. Draw the minute hand on the clock to show the correct time.

3:45   11:30   6:15

Lesson 13:   Construct a paper clock by partitioning a circle into halves and
quarters, and tell time to the half hour or quarter hour.

Name _____   Date _____

Draw the minute hand on the clock to show the correct time.

Half past 7   12:15   A quarter to 3

Lesson 13:   Construct a paper clock by partitioning a circle into halves and
              quarters, and tell time to the half hour or quarter hour.

© 2018 Great Minds®. eureka-math.org

79

R (Read the problem carefully.)

Brownies take 45 minutes to bake. Pizza takes half an hour less than brownies to warm up. How long does pizza take to warm up?

D (Draw a picture.)

W (Write and solve an equation.)

W (Write a statement that matches the story.)

_____

_____

_____

Lesson 14:   Tell time to the nearest five minutes.

EUREKA MATH

Name _____   Date _____

1. Fill in the missing numbers.

    60, 55, 50, _____, 40, _____, _____, _____, 20, _____, _____, _____, _____,

2. Fill in the missing numbers on the face of the clock to show the minutes.

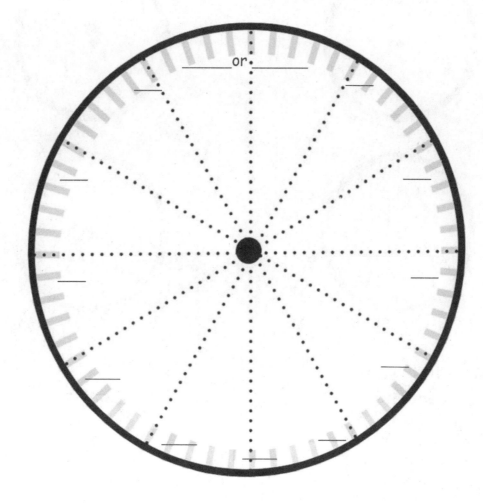

3. Draw the hour and minute hands on the clocks to match the correct time.

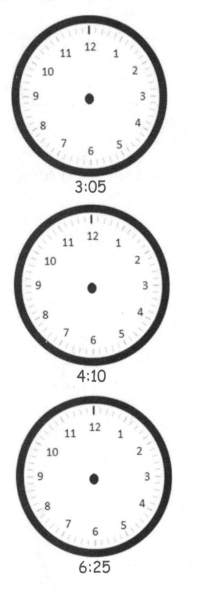

3:05　　　　　　　　　　　　3:35

4:10　　　　　　　　　　　　4:40

6:25　　　　　　　　　　　　6:55

4. What time is it?

_____　　　　　　_____

EUREKA MATH®

Name _____   Date _____

Draw the hour and minute hands on the clocks to match the correct time.

12:55

5:25

R (Read the problem carefully.)

At Memorial School, students have a quarter hour for morning recess and 33 minutes for a lunch break. How much free time do they have in all? How much more time for lunch than recess do they have?

D (Draw a picture.)

W (Write and solve an equation.)

Lesson 15:   Tell time to the nearest five minutes; relate *a.m.* and *p.m.* to time of day.

© 2018 Great Minds®. eureka-math.org

87

W (Write a statement that matches the story.)

_____

_____

_____

Lesson 15: Tell time to the nearest five minutes; relate *a.m.* and *p.m.* to time of day.

Name _____   Date _____

1. Decide whether the activity below would happen in the a.m. or the p.m. Circle your answer.

   a. Waking up for school   a.m. / p.m.

   b. Eating dinner   a.m. / p.m.

   c. Reading a bedtime story   a.m. / p.m.

   d. Making breakfast   a.m. / p.m.

   e. Having a play date after school a.m. / p.m.

   f. Going to bed   a.m. / p.m.

   g. Eating a piece of cake   a.m. / p.m.

   h. Eating lunch   a.m. / p.m.

  EUREKA
MATH

Lesson 15:   Tell time to the nearest five minutes; relate *a.m.* and *p.m.* to
               time of day.

© 2018 Great Minds®. eureka-math.org

89

2. Draw the hands on the analog clock to match the time on the digital clock. Then, circle **a.m. or p.m.** based on the description given.

   a. Brushing your teeth after you wake up

   a.m. or p.m.

   b. Finishing homework

   5:55   **a.m. or p.m.**

3. Write what you might be doing if it were **a.m. or p.m.**

   a. **a.m.** _____

   b. **p.m.** _____

4. What time does the clock show?

   _____ : _____

Lesson 15:   Tell time to the nearest five minutes; relate *a.m.* and *p.m.* to time of day.

© 2018 Great Minds®. eureka-math.org

Name _____   Date _____

Draw the hands on the analog clock to match the time on the digital clock. Then, circle **a.m. or p.m.** based on the description given.

1.   The sun is rising.

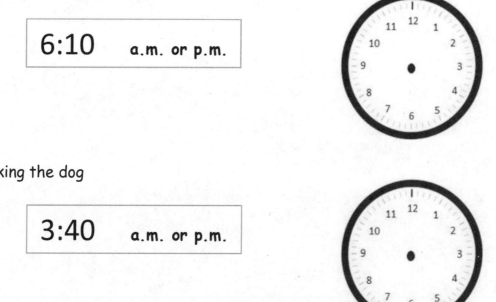

| 6:10 | a.m. or p.m. |

2.   Walking the dog

| 3:40 | a.m. or p.m. |

EUREKA MATH

Lesson 15:   Tell time to the nearest five minutes; relate *a.m.* and *p.m.* to
time of day.

91

© 2018 Great Minds®. eureka-math.org

Write the time.   Circle a.m. or p.m.

a.m./p.m.

telling time story (large)

Lesson 15:   Tell time to the nearest five minutes; relate *a.m.* and *p.m.* to time of day.

© 2018 Great Minds®. eureka-math.org

93

Write the time. Circle a.m. or p.m.

a.m./p.m.

telling time story (large)

Lesson 15:   Tell time to the nearest five minutes; relate *a.m.* and *p.m.* to
time of day.

Write the time.   Circle a.m. or p.m.

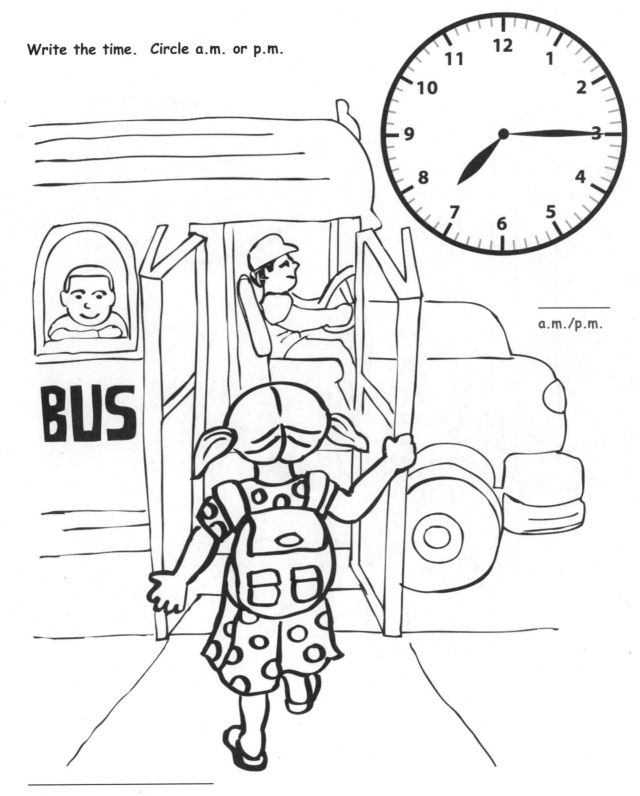

a.m./p.m.

telling time story (large)

Lesson 15:   Tell time to the nearest five minutes; relate *a.m.* and *p.m.* to time of day.

© 2018 Great Minds®. eureka-math.org

95

Write the time. Circle a.m. or p.m.

a.m./p.m.

telling time story (large)

Write the time. Circle a.m. or p.m.

a.m./p.m.

telling time story (large)

Lesson 15: Tell time to the nearest five minutes; relate *a.m.* and *p.m.* to
time of day.

© 2018 Great Minds®. eureka-math.org

97

Write the time. Circle a.m. or p.m.

_____

a.m./p.m.

telling time story (large)

Lesson 15:   Tell time to the nearest five minutes; relate *a.m.* and *p.m.* to time of day.

Write the time.   Circle a.m. or p.m.

_____
a.m./p.m.

_____

telling time story (large)

Lesson 15:   Tell time to the nearest five minutes; relate *a.m.* and *p.m.* to
               time of day.

© 2018 Great Minds®. eureka-math.org

99

Write the time. Circle a.m. or p.m.

_____
a.m./p.m.

_____
telling time story (large)

Lesson 15:   Tell time to the nearest five minutes; relate *a.m.* and *p.m.* to
time of day.

R (Read the problem carefully.)

On Saturdays, Jean may only watch cartoons for one hour. Her first cartoon lasts 14 minutes, and the second lasts 28 minutes. After a 5-minute break, Jean watches a 15-minute cartoon. How much time does Jean spend watching cartoons? Did she break her time limit?

D (Draw a picture.)

W (Write and solve an equation.)

Lesson 16:   Solve elapsed time problems involving whole hours and a half hour.

101

© 2018 Great Minds®. eureka-math.org

W (Write a statement that matches the story.)

_____

_____

_____

Name _____   Date _____

1. How much time has passed?

    a. 6:30 a.m. → 7:00 a.m.   _____

    b. 4:00 p.m. → 9:00 p.m.   _____

    c. 11:00 a.m. → 5:00 p.m.   _____

    d. 3:30 a.m. → 10:30 a.m.   _____

    e. 7:00 p.m. → 1:30 a.m.   _____

    f.

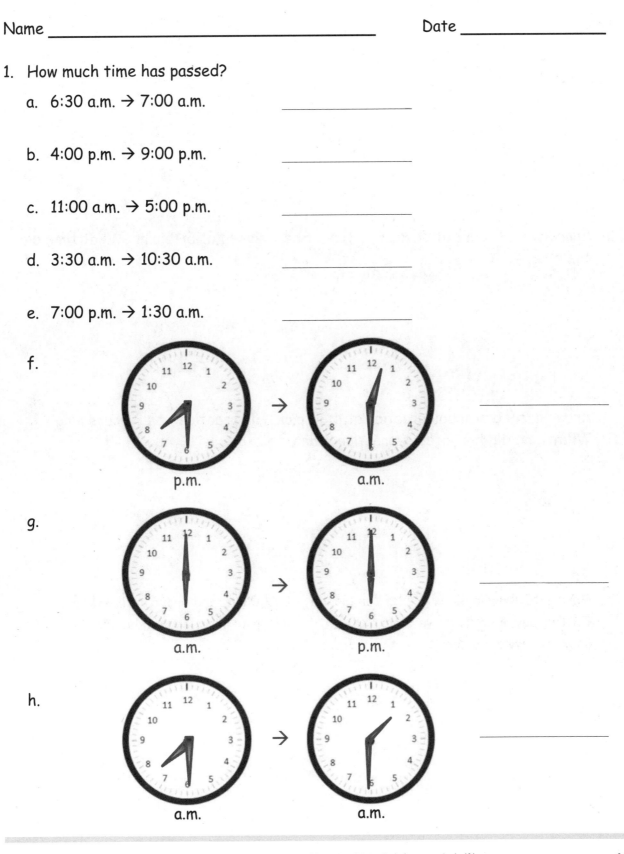

    g.

    h.

2. Solve.

    a. Tracy arrives at school at 7:30 a.m. She leaves school at 3:30 p.m. How long is Tracy at school?

    b. Anna spent 3 hours at dance practice. She finished at 6:15 p.m. What time did she start?

    c. Andy finished baseball practice at 4:30 p.m. His practice was 2 hours long. What time did his baseball practice start?

    d. Marcus took a road trip. He left on Monday at 7:00 a.m. and drove until 4:00 p.m. On Tuesday, Marcus drove from 6:00 a.m. to 3:30 p.m. How long did he drive on Monday and Tuesday?

Lesson 16:   Solve elapsed time problems involving whole hours and a half hour.

Name _____ Date _____

How much time has passed?

1. 3:00 p.m. → 11:00 p.m.   _____

2. 5:00 a.m. → 12:00 p.m. (noon)   _____

3. 9:30 p.m. → 7:30 a.m.   _____

Lesson 16:   Solve elapsed time problems involving whole hours and a half hour.

105

© 2018 Great Minds®. eureka-math.org

Credits

Great Minds® has made every effort to obtain permission for the reprinting of all copyrighted material. If any owner of copyrighted material is not acknowledged herein, please contact Great Minds for proper acknowledgment in all future editions and reprints of this module.